MATHadazzles Junior

Reasoning with Numbers

Volume 2

Authors

Carole Greenes

Mary Cavanagh

Contributors: Grades 2-3 Students
Shristi Bansal, Brianna Bautista, Grant Moberg,
Daphne Norman, Brent Robinson

Editors
Tanner Wolfram, Senior Editor
Jason Luc
Yifan Tian

Co-Leader
Joselyn Ly

Cover Design
Mary Cavanagh

Copyright © 2018 Arizona Board of Regents for and on behalf of
Arizona State University (ASU)

PRIME Center
Tempe, Arizona

MATHadazzles Junior Volume 2
Reasoning with Numbers

MATHadazzles are number puzzles that will develop your logical reasoning abilities, your sense of numbers, including computations with numbers, ways to compare numbers, and how to represent unknown numbers in equations and expressions.

What is a MATHadazzle? A MATHadazzle is a 2-by-3 grid with circles at the end of each row and at the bottom of each column. Some grid cells have clues about numbers that will fill those cells. Circles contain some of the sums of the numbers in that row or column.

What's Your Job? Based on clues provided in some of the grid cells, you place the numbers 5, 6, 7, 8, 9, and 10 in the six cells AND the given sums in the empty circles at the ends of the rows and at the bottoms of the columns. Sounds challenging? It is, but do not give up!

What are The Clues?

Computations with Numbers: The operations of addition, subtraction, multiplication and division with numbers, zero through 50, will be used in the clues. The result of each operation will be a number in the set of numbers, 5 – 10.

Ways to Compare Numbers: Numbers and expressions may be compared in three ways.
Greater than: The greater than symbol is >

Example: 10 is greater than 6	$10 > 6$
Example: 3 plus 4 is greater than 6	$3 + 4 > 6$

Greater than or equal to: The greater than or equal to symbol is ≥

Example: 7 is greater than or equal to 6	$7 \geq 6$
Example: 7 is greater than or equal to 7	$7 \geq 7$

Less than: The less than symbol is <

Example: 6 is less than 9 $5 < 9$
Example: 5 + 2 is less than 8 $5 + 2 < 8$

Less than or equal to: The less than or equal to symbol is $\leq$

Example: 4 is less than or equal to 9 $4 \leq 9$
Example: 4 is less than or equal to 4 $4 \leq 4$

Equal: The equal to symbol is =

Example: 9 is equal to 4 plus 5 $9 = 4 + 5$
Example: 10 minus 3 is equal to 7 $10 - 3 = 7$
Example: 2 times 3 is equal to 6 $2 \times 3 = 6$
Example: 10 divided by 2 is equal to 5 $10 \div 2 = 5$

Ways to Represent Unknown Numbers: All unknown numbers, in equations and expressions, will be represented by the letter *n*. The job of the solver is to figure out the value of *n*. In some problems *n* will have one value. In some problems *n* will have several values.

Equations with Unknowns
Example: $n + 5 = 12$, so $n = 7$.
Example: $16 - 10 = n$, so $n = 6$.
Example: $24 \div n = 3$, so $n = 8$.
Example: $10n = 60$, so $n = 6$.

Expressions with Unknowns
Example: $n > 8$, so in the set of numbers 5-10, *n* may be 9 or 10.
Example: $n < 7$, so in the set of numbers 5-10, *n* may be 5 or 6.
Example: $2 + n > 11$, so in the set of numbers 5-10, *n* must be 10.

Enjoy Solving!

Your MATHadazzling Authors, Contributors, and Editors

1

Put these numbers in the squares 5, 6, 7, 8, 9, 10

Add across →

Add down ↓

Sums are in ◯

Put these numbers in the circles: 11 17 24

$9 - 3 = n$		
$3 + 2 = n$		$4 + 5 = n$

Circles: 21, (empty), 17 below middle square, (empty) below left and right squares.

PRIME Center   MAThadazzles Junior Volume 2

2

Put these numbers in the squares 5, 6, 7, 8, 9, 10

Add across ⟶

Add down ↓

Sums are in ◯

Put these numbers in the circles: 13 15 24

$2+2+3=n$		$10-5=n$	21
		$6+1+3=n$	
	17		

PRIME Center   MATHadazzles Junior Volume 2

Put these numbers in the squares 5, 6, 7, 8, 9, 10

Add across →

Add down ↓

Sums are in ◯

Put these numbers in the circles: 16 18 20

		$5 + 5 = n$	25
	$10 - 8 + 5 = n$	$9 - 1 = n$	

11

4

Put these numbers in the squares 5, 6, 7, 8, 9, 10

Add across ⟶

Add down ↓

Sums are in ◯

Put these numbers in the circles: 13   22   23

		$1 + 7 = n$
	$60 \div 12 = n$	

Circles: 14, 18

5

Put these numbers in the squares 5, 6, 7, 8, 9, 10

Add across ⟶

Add down ↓

Sums are in ◯

Put these numbers in the circles: 13 17 21

	$64 \div n = 8$		(24)
	$n < 8$	$7 \times n = 49$	( )

(15)   ( )   ( )

PRIME Center   MATHadazzles Junior Volume 2

Put these numbers in the squares 5, 6, 7, 8, 9, 10

Add across ⟶

Add down ↓

Sums are in ◯

Put these numbers in the circles: 13 19 20

	$10 - n = 5$	$n - 1 = 8$	◯
	$4 \times 2 = n$		25
13	◯	◯	

7

Put these numbers in the squares 5, 6, 7, 8, 9, 10

Add across ⟶

Add down ↓

Sums are in ◯

Put these numbers in the circles: 11 15 19

	$7 \div 1 = n$	$10 - 1 = n$
$6 \div 1 = n$		

21

24

8

Put these numbers in the squares 5, 6, 7, 8, 9, 10

Add across →

Add down ↓

Sums are in ◯

Put these numbers in the circles: 13 13 20

		$n > 9$	25
	$n + 4 = 9$		◯
◯	◯	19	

PRIME Center MAThadazzles Junior Volume 2

9

Put these numbers in the squares 5, 6, 7, 8, 9, 10

Add across →

Add down ↓

Sums are in ◯

Put these numbers in the circles: 14 15 16

	$n > 8$	$80 \div n = 10$	23
$3n = 27$			22

PRIME Center   MATHadazzles Junior Volume 2

10

Put these numbers in the squares 5, 6, 7, 8, 9, 10

Add across ⟶

Add down ↓

Sums are in ◯

Put these numbers in the circles: 13   22   23

	$10 \div n = 2$	$3 + n = 10$

16   ◯   16

11

Put these numbers in the squares 5, 6, 7, 8, 9, 10

Add across ⟶

Add down ↓

Sums are in ◯

Put these numbers in the circles: 13 14 18

$n + n = 16$			◯ 21
	$4 \times 1 + 2 = n$	$n < 6$	◯ 21

◯ ◯ ◯

PRIME Center   MATHadazzles Junior Volume 2

12

Put these numbers in the squares 5, 6, 7, 8, 9, 10

Add across ⟶

Add down ↓

Sums are in ◯

Put these numbers in the circles: 16 16 18

$n + 2n = 30$		$3 \times 3 = n$

Circles: (right side, top) ◯, (right side, bottom) 27, (bottom left) ◯, (bottom middle) 13, (bottom right) ◯

13

Put these numbers in the squares 5, 6, 7, 8, 9, 10

Add across ⟶

Add down ↓

Sums are in ◯

Put these numbers in the circles: 15 15 21

$16 \div n = 2$	$3 \times 2 = n$	
		$10 \div n = 2$

Circles: 24, (empty), 15

PRIME Center   MATHadazzles Junior Volume 2

14

Put these numbers in the squares 5, 6, 7, 8, 9, 10

Add across ⟶

Add down ↓

Sums are in ◯

Put these numbers in the circles: 11 17 23

	$n > 6$	
$3 \times 2 + 1 = n$		$n > 5$

Circles: (right side, top) ◯, (right side, bottom) 22, (below left) 17, (below middle) ◯, (below right) ◯

15

Put these numbers in the squares 5, 6, 7, 8, 9, 10

Add across ⟶

Add down ↓

Sums are in ◯

Put these numbers in the circles: 14 14 17

	$2 + 4 + 1 = n$		**18**
$8 + 2 - 2 = n$			**27**

PRIME Center   MATHadazzles Junior Volume 2

16

Put these numbers in the squares 5, 6, 7, 8, 9, 10

Add across ⟶

Add down ↓

Sums are in ◯

Put these numbers in the circles: 13 17 21

	$10 - 4 = n$	
	$10 - 3 = n$	$4 + 5 = n$

◯ 24

◯

◯ 15   ◯   ◯

PRIME Center   MAThadazzles Junior Volume 2

17

Put these numbers in the squares 5, 6, 7, 8, 9, 10

Add across ⟶

Add down ↓

Sums are in ◯

Put these numbers in the circles: 15   16   22

	$60 \div 10 = n$	$14 \div 2 = n$
	$27 - n = 17$	

◯ 23

◯ 14   ◯   ◯

PRIME Center MATHadazzles Junior Volume 2

18

Put these numbers in the squares 5, 6, 7, 8, 9, 10

Add across →

Add down ↓

Sums are in ◯

Put these numbers in the circles: 13 15 27

	$3 + n = 9$	
	$n > 8$	$n + 10 = 20$

Circles: 18, (blank), 17

PRIME Center MAThadazzles Junior Volume 2

19

Put these numbers in the squares 5, 6, 7, 8, 9, 10

Add across ⟶

Add down ↓

Sums are in ◯

Put these numbers in the circles: 15 17 24

$48 \div n = 6$		$36 \div n = 6$

Circles: 21, 13

20

Put these numbers in the squares 5, 6, 7, 8, 9, 10

Add across ⟶

Add down ↓

Sums are in ◯

Put these numbers in the circles: 12 15 23

	$18 - n = 10$	$235 - n = 230$
$30 \div 5 = n$		

Circles: 22 (right of top row), ◯ (right of bottom row), ◯, 18, ◯ (below)

PRIME Center MATHadazzles Junior Volume 2

21

Put these numbers in the squares 5, 6, 7, 8, 9, 10

Add across →

Add down ↓

Sums are in ◯

Put these numbers in the circles: 12   14   23

$36 \div 9 + 2 = n$		$7 - 4 + 4 = n$
	$5 + 8 - 3 = n$	

Right circles (row sums): 22, ___

Bottom circles (column sums): ___, 19, ___

PRIME Center   MATHadazzles Junior Volume 2

22

Put these numbers in the squares 5, 6, 7, 8, 9, 10

Add across ⟶

Add down ↓

Sums are in ◯

Put these numbers in the circles: 15 15 23

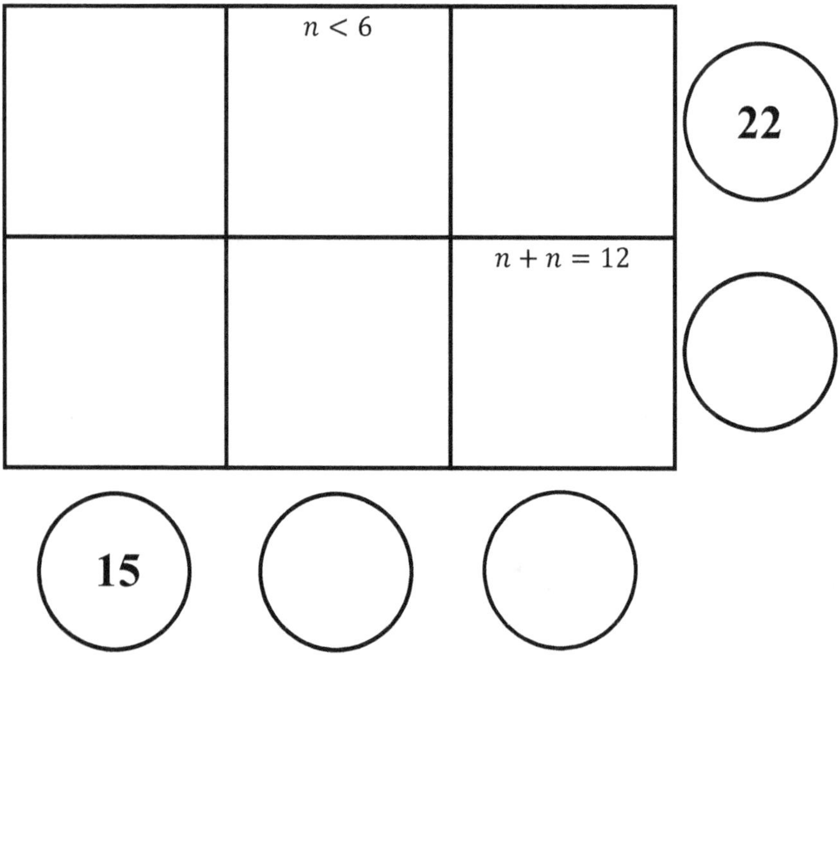

	$n < 6$	
		$n + n = 12$

Circles: 22, (blank), 15, (blank), (blank)

23

Put these numbers in the squares 5, 6, 7, 8, 9, 10

Add across ⟶

Add down ↓

Sums are in ◯

Put these numbers in the circles: 14 15 16

	$2 \times n = 14$	$30 \div n = 5$	⬤ 18
$7 \times n = 63$			⬤ 27

◯ ◯ ◯

PRIME Center   MATHadazzles Junior Volume 2

24

Put these numbers in the squares 5, 6, 7, 8, 9, 10

Add across ⟶

Add down ↓

Sums are in ◯

Put these numbers in the circles: 11 17 22

| | | $100 - 90 = n$ | | $10 \div 2 = n$ | 23 |

(17)

25

Put these numbers in the squares 5, 6, 7, 8, 9, 10

Add across ⟶

Add down ↓

Sums are in ◯

Put these numbers in the circles: 13   13   24

	$n \times 2 = 14$		21
		$7 + n = 15$	

19

PRIME Center   MATHadazzles Junior Volume 2

Put these numbers in the squares 5, 6, 7, 8, 9, 10

Add across ⟶

Add down ↓

Sums are in ◯

Put these numbers in the circles: 14 14 20

$25 \div n = 5$		$64 \div n = 8$

Circles: (top right) , (right middle) 25, (bottom left) , (bottom middle) 17, (bottom right)

27

Put these numbers in the squares 5, 6, 7, 8, 9, 10

Add across ⟶

Add down ↓

Sums are in ◯

Put these numbers in the circles: 14 17 18

$14 - n = 7$		$10 - n = 5$

Circle: 27

Circle: 14

PRIME Center   MATHadazzles Junior Volume 2

28

Put these numbers in the squares 5, 6, 7, 8, 9, 10

Add across ⟶

Add down ↓

Sums are in ◯

Put these numbers in the circles: 11   18   23

	$30 \div 5 = n$	$50 \div 5 = n$
$3 \times 3 = n$		

Circles: 16, (empty), (empty), 22, (empty)

PRIME Center   MATHadazzles Junior Volume 2

29

Put these numbers in the squares 5, 6, 7, 8, 9, 10

Add across ⟶

Add down ↓

Sums are in ◯

Put these numbers in the circles: 13   16   21

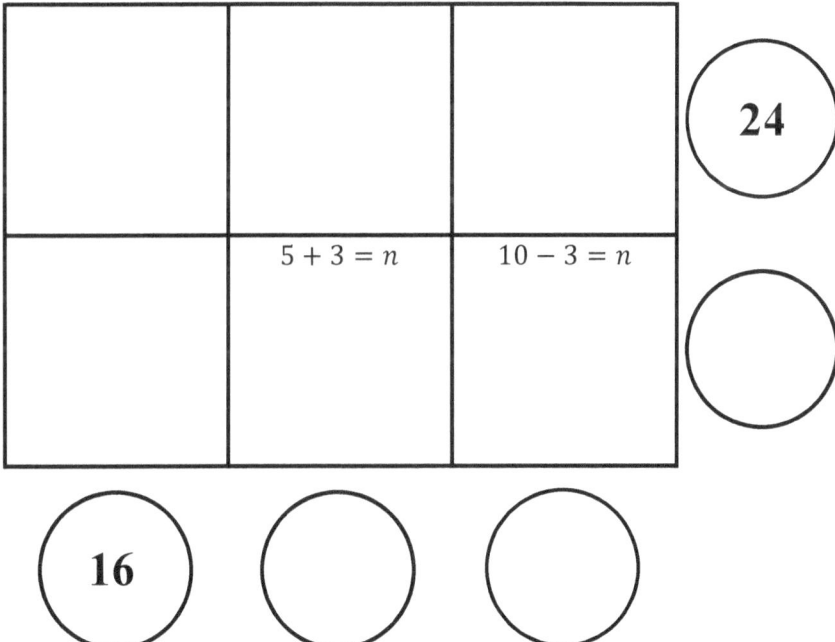

30

Put these numbers in the squares 5, 6, 7, 8, 9, 10

Add across →

Add down ↓

Sums are in ○

Put these numbers in the circles: 12 19 23

	$n + n = 10$		**22**
$n \times n = 100$		$n + n = 12$	
		14	

31

Put these numbers in the squares 5, 6, 7, 8, 9, 10

Add across ⟶

Add down ↓

Sums are in ◯

Put these numbers in the circles: 13   15   17

	$n \times 10 = 50$		ⓘ 18
$n > 9$			ⓘ 27

◯ ◯ ◯

PRIME Center   MATHadazzles Junior Volume 2

Put these numbers in the squares 5, 6, 7, 8, 9, 10

Add across ➔

Add down ↓

Sums are in ◯

Put these numbers in the circles: 12   15   18

	$n \geq 9$	
$n \times 2 = 18$		$n < 6$

23

22

Put these numbers in the squares 5, 6, 7, 8, 9, 10

Add across →

Add down ↓

Sums are in ◯

Put these numbers in the circles: 13 13 19

		$n < 8$
$n > 8$	$n + 3 = 9$	

22

23

Put these numbers in the squares 5, 6, 7, 8, 9, 10

Add across →

Add down ↓

Sums are in ◯

Put these numbers in the circles: 12 21 24

$10 - 5 = n$		
		$n < 7$

18   15

35

Put these numbers in the squares 5, 6, 7, 8, 9, 10

Add across ⟶

Add down ↓

Sums are in ◯

Put these numbers in the circles: 14 16 22

	$3 \times 5 - 6 = n$	$n < 6$	◯
	$n < 8$		23

◯ ◯ 15

PRIME Center MATHadazzles Junior Volume 2

36

Put these numbers in the squares 5, 6, 7, 8, 9, 10

Add across ⟶

Add down ↓

Sums are in ◯

Put these numbers in the circles: 12   15   18

7	8	9 ($n \times 2 = 18$)	(24)
5 ($n < 6$)	10	6	(21)
(12)	(18)	(15)	

PRIME Center

MATHadazzles Junior Volume 2

Put these numbers in the squares 5, 6, 7, 8, 9, 10

Add across ⟶

Add down ↓

Sums are in ◯

Put these numbers in the circles: 15   19   26

	$3 \times 3 = n$	
		$10 - 5 = n$

◯ ◯ 15 15

38

Put these numbers in the squares 5, 6, 7, 8, 9, 10

Add across ⟶

Add down ↓

Sums are in ◯

Put these numbers in the circles: 14 17 22

		$15 - n = 9$
$170 - n = 160$	$35 - n = 30$	

Circles: (top right) empty, (middle right) 23, (bottom left) empty, (bottom middle) empty, (bottom right) 14

39

Put these numbers in the squares 5, 6, 7, 8, 9, 10

Add across ⟶

Add down ↓

Sums are in ◯

Put these numbers in the circles: 15   19   24

		$150 - n = 140$
$14 \div 2 = n$	$25 - n = 20$	

Right circles: ◯ (top), **21** (bottom)

Bottom circles: ◯, **11**, ◯

PRIME Center   MATHadazzles Junior Volume 2

40

Put these numbers in the squares 5, 6, 7, 8, 9, 10

Add across →

Add down ↓

Sums are in ◯

Put these numbers in the circles: 13 19 21

		$100 \div n = 10$	◯ 24
$49 \div n = 7$			◯

◯   ◯ 13   ◯

PRIME Center   MAThadazzles Junior Volume 2

41

Put these numbers in the squares 5, 6, 7, 8, 9, 10

Add across ⟶

Add down ↓

Sums are in ◯

Put these numbers in the circles: 11   17   22

	$81 \div 9 = n$		◯
$100 - 90 = n$		$10 \div 2 = n$	23
◯	17	◯	

PRIME Center   MATHadazzles Junior Volume 2

42

Put these numbers in the squares 5, 6, 7, 8, 9, 10

Add across →

Add down ↓

Sums are in ◯

Put these numbers in the circles: 13 13 22

		$5 \times 2 = n$
$4 \times 2 = n$	$2 \times 3 = n$	

Right circles: (top) ◯, (bottom) 23

Bottom circles: ◯, ◯, 19

Put these numbers in the squares 5, 6, 7, 8, 9, 10

Add across ⟶

Add down ↓

Sums are in ◯

Put these numbers in the circles: 14   19   25

$25 \div 5 = n$		
		$64 \div 8 = n$

20

12

PRIME Center   MATHadazzles Junior Volume 2

Put these numbers in the squares 5, 6, 7, 8, 9, 10

Add across →

Add down ↓

Sums are in ◯

Put these numbers in the circles: 11   16   18

	$19 - 13 = n$	
		$3 + 7 - 3 = n$

23

22

45

Put these numbers in the squares 5, 6, 7, 8, 9, 10

Add across ⟶

Add down ↓

Sums are in ◯

Put these numbers in the circles: 11   19   23

$81 \div n = 9$		
		$6 \times n = 36$

◯ 22

◯

◯   ◯ 15   ◯

PRIME Center

Put these numbers in the squares 5, 6, 7, 8, 9, 10

Add across ⟶

Add down ↓

Sums are in ◯

Put these numbers in the circles: 11   16   18

$10 - n = 2$		
	$n < 6$	

21

24

47

Put these numbers in the squares 5, 6, 7, 8, 9, 10

Add across ⟶

Add down ↓

Sums are in ◯

Put these numbers in the circles: 11 17 24

$25 \div 5 = n$		$7 - 6 + 6 = n$
	$4 + 6 - 2 = n$	

◯ **21**

◯

◯ ◯ ◯ **17**

PRIME Center   MATHadazzles Junior Volume 2

48

Put these numbers in the squares 5, 6, 7, 8, 9, 10

Add across ⟶

Add down ↓

Sums are in ◯

Put these numbers in the circles: 11 19 21

$8 - 4 + 2 = n$			24
	$35 \div 5 = n$	$4 + 6 - 1 = n$	

15

PRIME Center MATHadazzles Junior Volume 2

49

Put these numbers in the squares 5, 6, 7, 8, 9, 10

Add across →

Add down ↓

Sums are in ◯

Put these numbers in the circles: 13   15   18

$n \times n = 25$		
		$18 - n = 8$

◯ (top right)

◯ 27

◯ ◯ ◯ 17

PRIME Center   MATHadazzles Junior Volume 2

50

Put these numbers in the squares 5, 6, 7, 8, 9, 10

Add across ⟶

Add down ↓

Sums are in ◯

Put these numbers in the circles: 18   22   23

$18 \div 3 = n$		$55 \div 5 - 2 = n$

13   ◯   14

51

Put these numbers in the squares 5, 6, 7, 8, 9, 10

Add across →

Add down ↓

Sums are in ○

Put these numbers in the circles: 11   18   21

6	$490 \div 70 = n$ (7)	8	(21)
5	9	$100 \div 10 = n$ (10)	24
(11)	16	(18)	

Put these numbers in the squares 5, 6, 7, 8, 9, 10

Add across ⟶

Add down ↓

Sums are in ◯

Put these numbers in the circles: 16 16 24

$4 \times 8 - 24 = n$		$30 \div n = 5$
	$90 \div n = 10$	

21

13

53

Put these numbers in the squares 5, 6, 7, 8, 9, 10

Add across ⟶

Add down ↓

Sums are in ◯

Put these numbers in the circles: 13   15   17

	$n > 5$	
$2 \times 4 = n$		$21 - n = 11$

◯ 20

◯ 25

◯   ◯   ◯

PRIME Center   MATHadazzles Junior Volume 2

Put these numbers in the squares 5, 6, 7, 8, 9, 10

Add across ⟶

Add down ↓

Sums are in ◯

Put these numbers in the circles: 13 15 17

	$n + n = 10$		18
		$n > 9$	27

◯ ◯ ◯

55

Put these numbers in the squares 5, 6, 7, 8, 9, 10

Add across ⟶

Add down ↓

Sums are in ◯

Put these numbers in the circles: 13 18 20

$3 \times 2 - 1 = n$		$10 \times 2 - n = 13$

◯

◯ 25

◯ 14   ◯   ◯

PRIME Center   MATHadazzles Junior Volume 2

Put these numbers in the squares 5, 6, 7, 8, 9, 10

Add across →

Add down ↓

Sums are in ◯

Put these numbers in the circles: 14 16 20

$49 \div n = 7$		$4 \times 2 = n$

25

15

Put these numbers in the squares 5, 6, 7, 8, 9, 10

Add across ⟶

Add down ↓

Sums are in ◯

Put these numbers in the circles: 15 15 27

	$81 \div n = 9$	
$5 - 1 + 1 = n$		$42 \div 6 = n$

Right circles: (top), 18

Bottom circles: 15, ◯, ◯

Put these numbers in the squares 5, 6, 7, 8, 9, 10

Add across ⟶

Add down ↓

Sums are in ◯

Put these numbers in the circles: 15 18 23

	$n + n = 10$	
$n + n = 20$		$7 > n$

Circles: 22, (empty), 12, and two empty circles.

Put these numbers in the squares 5, 6, 7, 8, 9, 10

Add across ⟶

Add down ↓

Sums are in ◯

Put these numbers in the circles: 13   15   24

$9 \times 9 - 72 = n$		$20 \div 4 = n$
	$5 + 8 - 3 = n$	

21

17

60

Put these numbers in the squares 5, 6, 7, 8, 9, 10

Add across ⟶

Add down ↓

Sums are in ◯

Put these numbers in the circles: 16 16 21

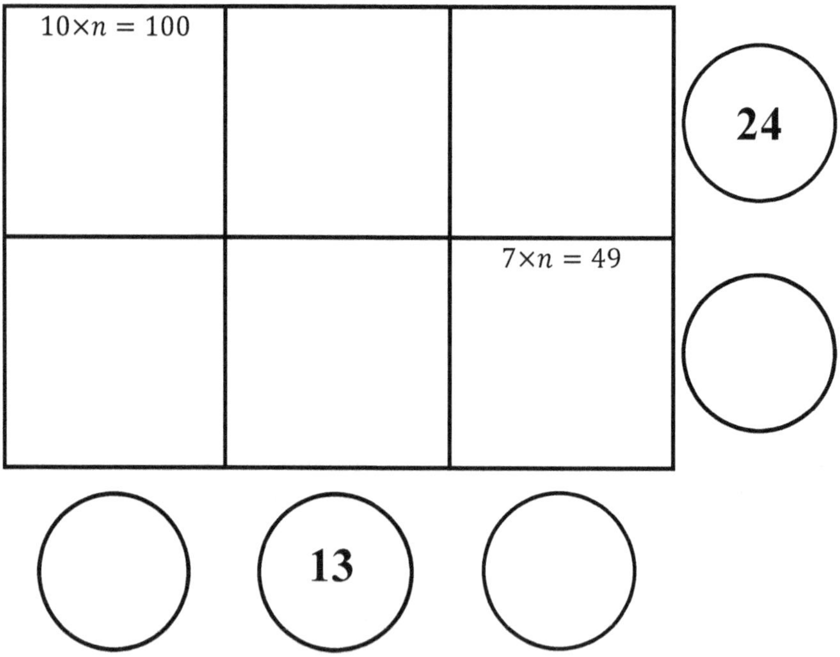

$10 \times n = 100$		
		$7 \times n = 49$

Circles: 24 (right of top row), empty (right of bottom row), empty, 13, empty (below columns)

61

Put these numbers in the squares 5, 6, 7, 8, 9, 10

Add across ⟶

Add down ↓

Sums are in ◯

Put these numbers in the circles: 11   19   21

$12 \div 2 = n$		$100 \div 10 = n$

Circles: (top right), 24, (bottom left), 15, (bottom right)

Put these numbers in the squares 5, 6, 7, 8, 9, 10

Add across ⟶

Add down ↓

Sums are in ◯

Put these numbers in the circles: 16 16 24

$36 \div n = 6$		
	$25 \div n = 5$	

21

13

63

Put these numbers in the squares 5, 6, 7, 8, 9, 10

Add across ⟶

Add down ↓

Sums are in ◯

Put these numbers in the circles: 15   17   21

$35 \div n = 5$		
		$81 \div n = 9$

Circles: (top right), 24 (middle right), (bottom left), 13 (bottom middle), (bottom right)

PRIME Center

MAThadazzles Junior Volume 2

Put these numbers in the squares 5, 6, 7, 8, 9, 10

Add across ⟶

Add down ↓

Sums are in ◯

Put these numbers in the circles: 13 17 18

	$12 - 6 = n$	$n > 6$

Circles on right: ◯, 27

Circles on bottom: ◯, 15, ◯

Put these numbers in the squares 5, 6, 7, 8, 9, 10

Add across →

Add down ↓

Sums are in ○

Put these numbers in the circles: 15 16 18

		$4 \times n = 32$
	$30 \div n = 5$	

27

14

66

Put these numbers in the squares 5, 6, 7, 8, 9, 10

Add across →

Add down ↓

Sums are in ◯

Put these numbers in the circles: 14   17   20

	$n < 7$	
$5 < n < 8$		$3 \times 4 - n = 2$

Circles: (top right) ◯ , (middle right) **25** , (bottom left) ◯ , (bottom middle) **14** , (bottom right) ◯

PRIME Center

MATHadazzles Junior Volume 2

Put these numbers in the squares 5, 6, 7, 8, 9, 10

Add across ⟶

Add down ↓

Sums are in ◯

Put these numbers in the circles: 14 15 16

$9 \times 3 - 18 = n$			22
	$5 < n < 9$		23

Put these numbers in the squares 5, 6, 7, 8, 9, 10

Add across ➝

Add down ↓

Sums are in ◯

Put these numbers in the circles: 14 18 23

$81 \div n = 9$		
		$100 \div n = 10$

Circles: (right side, top) ◯, (right side, bottom) 22

Bottom circles: ◯, 13, ◯

69

Put these numbers in the squares 5, 6, 7, 8, 9, 10

Add across →

Add down ↓

Sums are in ◯

Put these numbers in the circles: 16 22 23

$n < 6$			◯
	$n > 6$		◯
⓵④	⑮	◯	

(bottom circles: 14, 15, ◯)

Put these numbers in the squares 5, 6, 7, 8, 9, 10

Add across ⟶

Add down ↓

Sums are in ◯

Put these numbers in the circles: 14 16 25

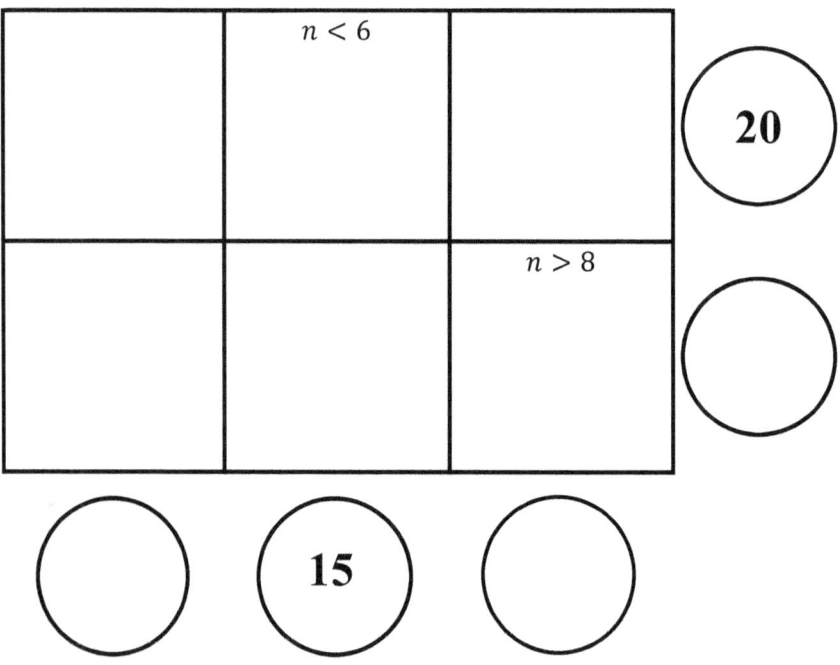

71

Put these numbers in the squares 5, 6, 7, 8, 9, 10

Add across ⟶

Add down ↓

Sums are in ◯

Put these numbers in the circles: 12   17   19

	$5 \times n = 50$		26
	$45 \div n = 9$		
	16		

PRIME Center   MATHadazzles Junior Volume 2

72

Put these numbers in the squares 5, 6, 7, 8, 9, 10

Add across ⟶

Add down ↓

Sums are in ◯

Put these numbers in the circles: 15 17 24

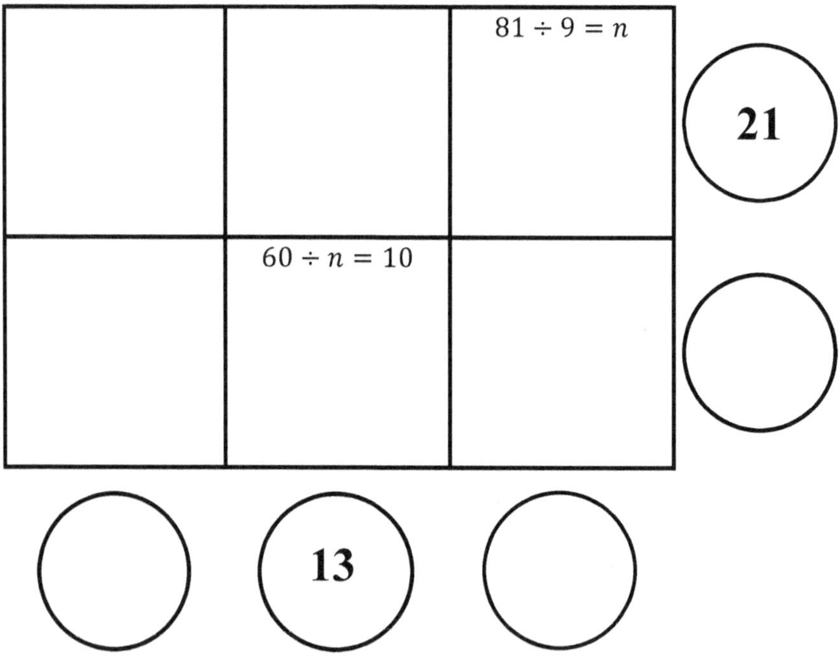

		$81 \div 9 = n$
	$60 \div n = 10$	

Circles: 21 (right top), (right bottom empty), (bottom left empty), 13 (bottom middle), (bottom right empty)

Put these numbers in the squares 5, 6, 7, 8, 9, 10

Add across ⟶

Add down ↓

Sums are in ◯

Put these numbers in the circles: 12   14   22

$1000 \div 100 = n$		$4 \times 9 - n = 28$	23
	$n > 6$		

19

Put these numbers in the squares 5, 6, 7, 8, 9, 10

Add across ⟶

Add down ↓

Sums are in ◯

Put these numbers in the circles: 11   16   18

$n + n = 16$		
		$n > 8$

20

25

75

Put these numbers in the squares 5, 6, 7, 8, 9, 10

Add across →

Add down ↓

Sums are in ◯

Put these numbers in the circles: 14 15 16

		$80 \div 10 = n$	⟵ 20
$n > 9$			⟵ 25

◯ ◯ ◯

Put these numbers in the squares 5, 6, 7, 8, 9, 10

Add across ⟶

Add down ↓

Sums are in ◯

Put these numbers in the circles: 13 22 23

	$n \times 2 = 10$	
		$4 + n = 13$

◯ 16   ◯   ◯ 16

PRIME Center   MATHadazzles Junior Volume 2

Put these numbers in the squares 5, 6, 7, 8, 9, 10

Add across →

Add down ↓

Sums are in ◯

Put these numbers in the circles: 13   15   17

$n \times 3 - 1 = 17$		$n > 8$
	$70 \div n = 7$	

20

25

Put these numbers in the squares 5, 6, 7, 8, 9, 10

Add across ⟶

Add down ↓

Sums are in ◯

Put these numbers in the circles: 11 16 22

		$25 \div n = 5$	◯
$81 \div 9 = n$			◯ 23
◯	◯ 18	◯	

Answers

1.

6	7	8	㉑
5	10	9	㉔

⑪ ⑰ ⑰

2.

7	9	5	㉑
6	8	10	㉔

⑬ ⑰ ⑮

3.

6	9	10	㉕
5	7	8	⑳

⑪ ⑯ ⑱

4.

6	9	8	㉓
7	5	10	㉒

⑬ ⑭ ⑱

5.

6	8	10	㉔
9	5	7	㉑

⑮ ⑬ ⑰

6.

6	5	9	⑳
7	8	10	㉕

⑬ ⑬ ⑲

Answers

7.

5	7	9	㉑
6	8	10	㉔
⑪	⑮	⑲	

8.

7	8	10	㉕
6	5	9	⑳
⑬	⑬	⑲	

9.

5	10	8	㉓
9	6	7	㉒
⑭	⑯	⑮	

10.

10	5	7	㉒
6	8	9	㉓
⑯	⑬	⑯	

11.

8	7	9	㉑
10	6	5	㉑
⑱	⑬	⑭	

12.

6	5	7	⑱
10	8	9	㉗
⑯	⑬	⑯	

Answers

13.

8	6	10	(24)
7	9	5	(21)
(15)	(15)	(15)	

14.

10	8	5	(23)
7	9	6	(22)
(17)	(17)	(11)	

15.

6	7	5	(18)
8	10	9	(27)
(14)	(17)	(14)	

16.

10	6	8	(24)
5	7	9	(21)
(15)	(13)	(17)	

17.

9	6	7	(22)
5	10	8	(23)
(14)	(16)	(15)	

18.

5	6	7	(18)
8	9	10	(27)
(13)	(15)	(17)	

Answers

19.

8	10	6	(24)
5	7	9	(21)
(13)	(17)	(15)	

20.

9	8	5	(22)
6	10	7	(23)
(15)	(18)	(12)	

21.

6	9	7	(22)
8	10	5	(23)
(14)	(19)	(12)	

22.

8	5	9	(22)
7	10	6	(23)
(15)	(15)	(15)	

23.

5	7	6	(18)
9	8	10	(27)
(14)	(15)	(16)	

24.

7	9	6	(22)
10	8	5	(23)
(17)	(17)	(11)	

PRIME Center MATHadazzles Junior Volume 2

Answers

25.

9	7	5	(21)
10	6	8	(24)
(19)	(13)	(13)	

26.

5	7	8	(20)
9	10	6	(25)
(14)	(17)	(14)	

27.

7	6	5	(18)
10	8	9	(27)
(17)	(14)	(14)	

28.

7	6	10	(23)
9	5	8	(22)
(16)	(11)	(18)	

29.

10	5	9	(24)
6	8	7	(21)
(16)	(13)	(16)	

30.

9	5	8	(22)
10	7	6	(23)
(19)	(12)	(14)	

Answers

31.

7	5	6	(18)
10	8	9	(27)
(17)	(13)	(15)	

32.

6	10	7	(23)
9	8	5	(22)
(15)	(18)	(12)	

33.

10	7	5	(22)
9	6	8	(23)
(19)	(13)	(13)	

34.

5	10	9	(24)
7	8	6	(21)
(12)	(18)	(15)	

35.

8	9	5	(22)
6	7	10	(23)
(14)	(16)	(15)	

36.

7	8	9	(24)
5	10	6	(21)
(12)	(18)	(15)	

Answers

37.

7	9	10	(26)
8	6	5	(19)
(15)	(15)	(15)	

38.

7	9	6	(22)
10	5	8	(23)
(17)	(14)	(14)	

39.

8	6	10	(24)
7	5	9	(21)
(15)	(11)	(19)	

40.

6	8	10	(24)
7	5	9	(21)
(13)	(13)	(19)	

41.

7	9	6	(22)
10	8	5	(23)
(17)	(17)	(11)	

42.

5	7	10	(22)
8	6	9	(23)
(13)	(13)	(19)	

Answers

43.

5	9	6	(20)
7	10	8	(25)
(12)	(19)	(14)	

44.

8	6	9	(23)
10	5	7	(22)
(18)	(11)	(16)	

45.

9	8	5	(22)
10	7	6	(23)
(19)	(15)	(11)	

46.

8	6	7	(21)
10	5	9	(24)
(18)	(11)	(16)	

47.

5	9	7	(21)
6	8	10	(24)
(11)	(17)	(17)	

48.

6	8	10	(24)
5	7	9	(21)
(11)	(15)	(19)	

Answers

49.

5	6	7	(18)
8	9	10	(27)
(13)	(15)	(17)	

50.

6	8	9	(23)
7	10	5	(22)
(13)	(18)	(14)	

51.

6	7	8	(21)
5	9	10	(24)
(11)	(16)	(18)	

52.

8	7	6	(21)
5	9	10	(24)
(13)	(16)	(16)	

53.

9	6	5	(20)
8	7	10	(25)
(17)	(13)	(15)	

54.

6	5	7	(18)
9	8	10	(27)
(15)	(13)	(17)	

Answers

55.

5	8	7	(20)
9	10	6	(25)
(14)	(18)	(13)	

56.

7	10	8	(25)
9	5	6	(20)
(16)	(15)	(14)	

57.

10	9	8	(27)
5	6	7	(18)
(15)	(15)	(15)	

58.

8	5	9	(22)
10	7	6	(23)
(18)	(12)	(15)	

59.

9	7	5	(21)
6	10	8	(24)
(15)	(17)	(13)	

60.

10	5	9	(24)
6	8	7	(21)
(16)	(13)	(16)	

Answers

61.

5	7	9	(21)
6	8	10	(24)
(11)	(15)	(19)	

62.

6	8	7	(21)
10	5	9	(24)
(16)	(13)	(16)	

63.

7	8	6	(21)
10	5	9	(24)
(17)	(13)	(15)	

64.

5	6	7	(18)
8	9	10	(27)
(13)	(15)	(17)	

65.

9	10	8	(27)
5	6	7	(18)
(14)	(16)	(15)	

66.

8	5	7	(20)
6	9	10	(25)
(14)	(14)	(17)	

PRIME Center

MATHadazzles Junior Volume 2

Answers

67.

9	8	5	(22)
7	6	10	(23)
(16)	(14)	(15)	

68.

9	6	8	(23)
5	7	10	(22)
(14)	(13)	(18)	

69.

5	8	10	(23)
9	7	6	(22)
(14)	(15)	(16)	

70.

8	5	7	(20)
6	10	9	(25)
(14)	(15)	(16)	

71.

9	10	7	(26)
8	6	5	(19)
(17)	(16)	(12)	

72.

5	7	9	(21)
10	6	8	(24)
(15)	(13)	(17)	

Answers

73.

10	5	8	㉓
9	7	6	㉒
⑲	⑫	⑭	

74.

8	5	7	⑳
10	6	9	㉕
⑱	⑪	⑯	

75.

5	7	8	⑳
10	9	6	㉕
⑮	⑯	⑭	

76.

10	5	7	㉒
6	8	9	㉓
⑯	⑬	⑯	

77.

6	5	9	⑳
7	10	8	㉕
⑬	⑮	⑰	

78.

7	10	5	㉒
9	8	6	㉓
⑯	⑱	⑪	

PRIME Center — MATHadazzles Junior Volume 2

A new Junior Series!
MATHadazzles Junior

- **Volume 1 Reasoning with Numbers**
- **Volume 2 Reasoning with Numbers**

Contributors to *MATHadazzles Junior* Volumes 1 and 2 are 2nd and 3rd grade students from the Greater Phoenix area.

The MATHadazzles Series includes:

- **Volume 1 Reasoning with Numbers**
- **Volume 2 Reasoning with Whole Numbers**
- **Volume 3 Reasoning with Integers**
- **Volume 4 Reasoning with Fractions**
- **Volume 5 Reasoning with Decimals**
- **Volume 6 Reasoning Algebraically**
- **Volume 7 Reasoning Algebraically**
- **Volume 8 Reasoning Algebraically with Decimals**

Contributors to Volumes 1, 2, and 3 are middle school teachers in the Greater Phoenix area.

Contributors to Volumes 4 and 5 are middle school students from the Greater Phoenix area.

Contributors to Volumes 6, 7 and 8 are high school students from the Greater Phoenix area who participated in the NSF Project AMP (#1509105).

www.ingramcontent.com/pod-product-compliance
Lightning Source LLC
Chambersburg PA
CBHW070310230526
45470CB00002B/804